COLE'S COOKING COMPANION SERIES

THE RICE COOKER

COLE GROUP

Both U.S. and metric units are provided for all recipes in this book. Ingredients are listed with U.S. units on the left and metric units on the right. The metric quantities have been rounded for ease of use; as a result, in some recipes there may be a slight difference (approximately ½ ounce or 15 grams) between the portion sizes for the two types of measurements.

Front cover photograph: Kevin Sanchez
Page 9: Joyce Chen Products—"Good Earth" Steam Pot
Page 11: Zojirushi America Corporation

Cole Group, Inc.
1330 N. Dutton Ave., Suite 103
Santa Rosa, CA 95401
(800) 959-2717 (707) 526-2682
Fax (707) 526-2687

Printed in Hong Kong

G F E D C B A
1 0 9 8 7 6 5

ISBN 1-56426-810-1

Library of Congress Catalog Card Number 95-16392

Distributed to the book trade by Publishers Group West

Contents

Getting Started 5

The Long and Short of It 6

Storing and Preparing Rice 8

Cooking Rice Right 8

Recipes and Techniques for The Rice Cooker 13

Appetizers & Hors d'Oeuvres 14

Nigiri Sushi 21

Neverfail Rice 22

Starters 26

Four-Seasons Rice Salad 35

Off-the-Shelf Rice Salad 39

Main Courses 40

Rice for Risotto 52

Preparing "Fried Rice" 61

The Stir-Fry Pantry 67

Accompaniments 70

Dry-Roasting Rice 76

Preparing Rice-Based Sopa Secas 80

Preparing Risotto 85

Side Dishes for All Seasons 86

Something Sweet 90

Index 95

Getting Started

With a soothing texture and subtle, nutty flavor you never tire of, rice blends comfortably and graciously into almost any menu. A fragrant bed of perfectly cooked rice provides a pleasing background for kabobs, curries, and a myriad of sauces. Artfully combined with meats, poultry, fish, legumes, vegetables or fruits, rice transforms itself into pilaf, paella, risotto, stir-fry, and dozens of other rice dishes to fit every occasion.

The Long and Short of It

Inexpensive, easy to store, impressively high in complex carbohydrates and other nutrients, infinitely versatile, and above all, delicious, rice is the staple fare for half the planet and the grain of choice throughout much of the other half. Besides the thousands of different varieties cultivated worldwide, commercial processing techniques have further increased the selection of rice products available. The amazing diversity of products displayed in the rice section of any well-stocked supermarket, natural food store, or gourmet food shop reflects the universal appeal of rice.

In the Grain

The most important consideration in selecting rice—at least from the cook's perspective—may well be the length of the grain. *Long-grain varieties* cook up into separate, dry, light grains, ideal for salads and pilafs. The fragrance and flavor of aromatic long-grain varieties such as basmati and a related variety, wehani, have made them the preferred rice for Indian cooking and a favorite all-purpose rice among Westerners. Even longer than the longest-grain variety of tropical rice is a distant relative—wild rice—the only "rice" that is native to the northern United States and Canada.

Medium-grain varieties such as Calrose are moist and slightly sticky when cooked—qualities that are highly desirable for dishes such as croquettes, puddings, and sushi. *Short-grain varieties*—including Arborio, the prime ingredient in any authentic Italian risotto—are good choices for puddings and other creamy-textured rice dishes. Glutinous rice, the stickiest of any short-grain variety (see page 23), is used in molded puddings and Japanese mochi, a chewy rice cake made from pounded cooked rice.

Through the Mill

In addition to the length of the grain, another important consideration for most cooks is the amount of processing rice receives before reaching the market. The least processed is *brown rice,* with only the hull removed; virtually all the nutrients remain intact. Brown rice requires longer cooking than *white rice,* which has its bran and other nutrient-rich components removed to extend shelf life and produce a pearlescent appearance. Both brown and white rice can be ground into *rice flour* and used in baked goods. *Converted rice* undergoes parboiling before it is milled to loosen the hull and diffuse nutrients from the bran into the rice kernel. *Enriched rice* has a coating of vitamin solution and protein powder applied to it in an attempt to replace nutrients lost during milling. Rice that has been precooked and dried is marketed as *quick-cooking,* while *flavored rice* has various seasonings and dehydrated vegetables or vermicelli added to the rice.

Storing and Preparing Rice

Rice that has been properly harvested, processed, and stored will keep indefinitely. Like all grains, rice (including packaged rice products) keeps best in airtight containers stored in a cool, dry place.

To prepare rice for cooking, many cooks like to first pick it over and remove any bits of chaff or other debris, then wash the grain (by swirling it vigorously in several changes of cool water) and drain it. While this process removes milling dust and seems to give the grain a pure, clean taste, it can also leach out nutrients, especially if the grain soaks for any length of time. Whether the nutritional loss is worth the improvement in purity and flavor is a matter of personal opinion.

Cooking Rice Right

Good cooks the world over are adamant about the "right" way to cook rice. In *The Rice Cooker* you'll find a half-dozen or so of the most widely used methods for cooking plain rice as well as techniques for preparing risotto, pilaf, and other classic rice dishes. Undoubtedly there are yet other ways of cooking rice. Given the many varieties of rice and the range of cooking methods to choose from, it makes sense to try several different ways until you discover which ones consistently yield rice cooked the way you like it. *That's* the right way to cook rice.

Other than a heat source, the only equipment you need to cook rice is a cooking container, and every culture that reveres rice has its favorites—from clay pots to electric steamers (see photos on pages 9 and 11). As to whether to salt the cooking water for rice, it's a matter of taste. In Chinese and other Asian cuisines, rice is typically cooked without salt. If you prefer salted rice, add salt to the cooking water (½ teaspoon for each cup or 250 ml of uncooked rice). In deciding how much to cook, keep in mind that uncooked rice generally triples in

volume as it cooks. For meals in which rice is served plain, allow ⅓–⅔ cup (85–150 ml), uncooked rice per person.

The cooking times given in the following guidelines are approximate; the age of the rice, humidity, altitude, and other variables all affect the cooking process.

Measured Water Method In this technique the rice is cooked, covered, in a measured quantity of water until all the liquid has been absorbed, producing moist kernels that retain all their nutrients.

White rice To 1½ cups (350 ml) boiling water add 1 cup (250 ml) rice and stir once with a fork. Cover, reduce heat to low, and cook 18 minutes. Remove from heat and let stand 5 minutes. Fluff with a fork.

Brown rice Follow instructions for white rice but increase water to 2 cups (500 ml) and cook 45–50 minutes.

Wild rice Follow instructions for white rice but increase water to 2½ cups (600 ml) and cook 40–45 minutes.

Chinese Method White rice works best for this technique, in which the grain is stirred during cooking to produce a slightly sticky rice that is preferred for many Chinese-style dishes: To 2 cups (500 ml) cold water add 1 cup (250 ml) rice. Bring to a boil over high heat, cover, and cook until water has been absorbed down to the level of the rice. Stir well, cover, and reduce heat to low; cook until grains are tender (about 15 minutes). Fluff with a fork.

Unlimited Water Method Similar to the technique used for cooking pasta, this method calls for boiling rice in a large quantity of water, then pouring off the excess water and completing the cooking over steam or in an oven set at a low temperature. Although any water-soluble nutrients go down the drain when the cooking liquid is discarded, this method requires no measuring and yields fluffy, dry grains that are appreciated in many types of Asian cooking.

White rice Boil rice, uncovered, in a large amount of water for 12–15 minutes until somewhat softened but still firm in the center. Drain well. Transfer to a cheesecloth-lined steamer and steam over simmering water until fluffy and dry, or place in a shallow baking dish, cover, and bake in a 300°F (150°C) oven for 10–15 minutes. Fluff with a fork.

Brown rice Follow directions for white rice but boil the rice for 25–30 minutes before steaming or baking.

Microwave Method Cooking rice in the microwave frees up the stovetop, saves on cleanup, and keeps the kitchen cool. Microwaving rice generally isn't any faster than conventional cooking methods, unless you use a steam pressure container specially designed for use in a microwave oven. Any cookware you use should be microwave-safe and large enough to prevent spillovers when the rice boils. Always follow the manufacturer's recommendations for operating any microwave oven.

White rice Combine 1 cup (250 ml) rice with 2 cups (500 ml) water, cover, and microwave 8 minutes at full power. Remove from microwave oven and allow to stand 10–15 minutes. Fluff with a fork.

Brown rice Combine 1 cup (250 ml) rice with 2½ cups (600 ml) water, cover, and microwave 20–25 minutes at full power. Remove from microwave oven and allow to stand 15–20 minutes. Fluff with a fork.

Electric Rice Cooker Method Most electric rice cookers, from the simplest to the most sophisticated state-of-the-art models, operate on a similar principle: You place a measured quantity of rice and water into the liner pan, set it over the heating element in the bottom of the rice cooker, and activate the machine. The rest is automatic. Some models with computerized controls can be programmed to cook brown or glutinous rice in addition to white rice, as well as paella and other rice specialties. For specific instructions, follow the manufacturer's recommendations.

Recipes and Techniques for The Rice Cooker

*I*ts subtle aroma, delicate flavor, and pleasing texture have earned rice a place of honor in cuisines all over the world. From paella to pilau, kedgeree to torta riso, rice assumes a myriad of forms—all of them delightful. In addition to more than 50 delicious recipes, *The Rice Cooker* includes methods for cooking specific rice varieties and techniques for preparing authentic sushi, risotto, pilaf, and other traditional and contemporary rice dishes.

Appetizers & Hors d'Oeuvres

If you're looking for out-of-this world canapés for a dinner party or snacks for a Sunday afternoon of videos, you'll find them in the following section. Stuffed into grape leaves for dolmas, rolled into sushi, or fried into golden croquettes with molten cheese centers, rice makes many a savory appetizer and hors d'oeuvre.

Pearl Balls with Pine Nut-Spinach Filling

These rice-studded appetizers make a good first course.

Pine Nut-Spinach Filling

½ lb	fresh spinach	225 g
1 tsp	soy sauce	1 tsp
½ tsp	sugar	½ tsp
1 tsp	Asian sesame oil	1 tsp
2 tbl	toasted pine nuts	2 tbl

¾ cup	glutinous rice, washed and drained	175 ml
¼ cup	white long-grain rice, washed and drained	60 ml
1 lb	lean ground pork	450 g
¼ lb	shrimp, peeled, deveined, and chopped	115 g
¼ cup	minced, peeled water chestnuts	60 ml
2	green onions, white part only, minced	2
1 tsp	grated fresh ginger	1 tsp
½ tsp each	salt and sugar	½ tsp each
⅛ tsp	white pepper	⅛ tsp
as needed	soy sauce and Asian sesame oil	as needed
1 tbl	dry vermouth or rice wine	1 tbl
1	egg, slightly beaten	1
1 tbl	cornstarch	1 tbl
as needed	spinach leaves, for lining steamer	as needed

1. To prepare filling, blanch spinach and squeeze out excess water; chop coarsely. Transfer to a small bowl and mix with soy sauce, sugar, sesame oil, and pine nuts. Refrigerate until ready to use.

2. Combine and soak both kinds of rice in enough cold water to cover for 4 hours or overnight. Drain well and set aside.

3. In a medium bowl combine pork, shrimp, water chestnuts, green onions, ginger, salt, sugar, white pepper, 1 tablespoon

soy sauce, 1 teaspoon sesame oil, and vermouth. Add egg and blend well. Sprinkle with cornstarch and mix together.

4. With moistened hands take a heaping tablespoon of pork mixture and flatten with hand. Place ½ teaspoon spinach filling in center of pork mixture and mold into densely packed 1½-inch-diameter (3.75-cm) ball. Continue with remaining mixture. Set aside.

5. Line steamer with 1 layer of spinach leaves. Spoon soaked rice into a shallow dish. Roll pork balls in rice and pat gently to embed grains. Arrange balls ½ inch (1.25 cm) apart on spinach.

6. Cover and steam over medium-high heat 20 minutes. Serve hot with soy sauce and sesame oil for dipping.

Makes about 12 balls.

Sizzling Rice Cakes

Serve these deep-fried cakes with hot soup (see page 29) or dips.

1 cup	**long-grain rice, washed and drained**	**250 ml**
1–2 tbl	**oil**	**1–2 tbl**
as needed	**peanut oil, for deep-frying**	**as needed**

1. In a heavy saucepan over high heat, bring rice and 2 cups (500 ml) water to a boil. Cover pan, reduce heat to low, and simmer 20 minutes, or until all water has been absorbed. Do not stir.

2. Drizzle oil over surface of rice. Loosen rice from pan and cut into cakes about 2 by 3 inches (5 by 7.5 cm).

3. Pour oil into a skillet to a depth of ½ inch (1.25 cm). Heat until a grain of rice dropped in oil sizzles. Fry rice cakes until light golden brown, then return cakes to hot oil for final crisping just before serving. Drain and serve hot.

Makes about 8 rice cakes.

California Roll Sushi

Sushi is a popular Japanese snack of vinegared rice rolled around a variety of ingredients such as seafood and fresh or pickled vegetables. West Coast sushi chefs combine crab and avocado with wasabi (green Japanese horseradish) and nori (dried seaweed) to create a popular California specialty.

½ tsp	wasabi powder	½ tsp
1	avocado, ripe but still firm	1
2 sheets	nori	2 sheets
1 recipe	Sushi Rice, including reserved vinegar mixture (see page 20)	1 recipe
½ cup	cooked crabmeat	125 ml

1. Combine wasabi with just enough water to form a loose paste and set aside. Split avocado, remove the pit and peel, cut flesh lengthwise into ¼-inch (.6-cm) sticks, and set aside. Toast nori by using tongs to hold each sheet over a gas flame for a few seconds, or until color turns from black to dark green.

2. Place nori on a bamboo rolling mat or a square of foil. Gently spread the rice ¼ inch (.6 cm) thick over nori, leaving ½-inch (1.25-cm) borders. Dab a very small amount of wasabi paste along length of rice. Lay half the avocado strips end to end along length of roll, then lay half of the crab pieces alongside avocado strips.

3. Moisten hands with vinegar mixture reserved from making Sushi Rice. Lifting bamboo mat or foil from edge nearest you, firmly roll nori and rice over fillings, enclosing them and forming a roll. Repeat with remaining ingredients.

4. Store rolls, loosely covered, for up to 2 hours before serving. To serve, uncover and slice into 1-inch-long (2.5-cm-long) sections. Arrange, cut side up, on plates or serving tray.

Serves 4.

SUSHI RICE

Use this sushi rice mixture to make California Roll Sushi (see page 18) or Nigiri Sushi (see page 21). Any leftover sushi rice can be formed into small balls, rolled in black sesame seeds, and served as an appetizer with soy sauce and wasabi paste.

2 cups	**short-grain rice, washed and drained**	**500 ml**
½ cup	**rice vinegar**	**125 ml**
¼ cup	**sugar**	**60 ml**
4 tsp	**salt**	**4 tsp**

1. Place rice in pot, cover with 2½ cups (600 ml) water, and soak 1 hour. Cover pot and bring to a boil. When steam escapes from around lid, reduce heat to low and simmer 20 minutes. Remove from heat and let stand 10 minutes longer.
2. While rice is cooking, combine vinegar, sugar, and salt in a small saucepan and cook until sugar dissolves. Cool.
3. Using a wooden spatula, scoop cooked rice into a shallow baking pan. Sprinkle with half of the vinegar mixture. Fold liquid in carefully without mashing grains. Using a folded newspaper, fan rice while folding in liquid. Continue fanning and folding until liquid is nearly all absorbed and rice glistens. Reserve remaining vinegar mixture for moistening hands while forming sushi.

Makes 4 cups (900 ml).

Nigiri Sushi

Any home cook who can prepare Sushi Rice and wield a sharp knife can make a fancy platter of nigiri sushi—appetizer-sized ovals of rice topped with the freshest of uncooked fish or shellfish. The paper-thin slices of tuna and butterflied shrimp shown here are particularly flavorful and eye-catching. Serve nigiri sushi at once, before it loses the freshness that makes it so succulent.

1. With a wooden spatula held vertically, cut vinegar mixture (see recipe on opposite page) into cooked rice, being careful not to mash the rice grains. Mix until rice is glossy and holds together but is not mushy.

2. Moisten hands with vinegar mixture and form a small amount of rice into an oval.

3. Top the oval of sushi rice with a dab of wasabi paste and a paper-thin slice of tuna or a butterflied shrimp.

Neverfail Rice

This technique, a variation on the Measured Water Method (see page 9), produces perfectly "steamed" white or brown rice, using the principle of absorption. Glutinous (sweet or sticky) rice requires a slightly different method (see page 23).

1. Measure the rice, allowing ⅓–⅔ cup (85–150 ml) uncooked rice per person, into a large bowl or pot. Add cold water to cover the rice by several inches and swirl the water vigorously until it turns cloudy.

2. Drain rice through a strainer. Repeat the swirling and straining procedure until water is clear, 3–5 times in all. Drain rice thoroughly and place in cooking pot. Cover rice with cold water 1 inch (2.5 cm)—the length of one joint of a finger or thumb—deeper than the depth of the rice. If you prefer more precise measurements, use the following guidelines to find the proper ratio of water for the type of rice to be cooked:

- Long-grain rice: Use 2 cups (500 ml) water for the first cup (250 ml) of rice and 1 cup (250 ml) of water for each additional cup (250 ml).
- Medium- or short-grain rice: Use 1½ cups (350 ml) water for the first cup (250 ml) of rice and 1 cup (250 ml) of water for each additional cup (250 ml).

3. Bring water to a boil, cover, reduce heat, and simmer until water is absorbed. Short- or medium-grain white rice will take about 15 minutes; short- or medium-grain brown rice will take 25–30 minutes. Long-grain rice (white or brown) cooks more quickly than short- or medium-grain, so time it accordingly. Lift the cover just long enough to check whether the rice has absorbed the water, then quickly replace the cover. When all the water is absorbed, turn off the heat and let the rice stand, covered, 10–15 minutes. Fluff with a fork or chopsticks before serving.

For glutinous rice, use about ⅓ cup (85 ml) of uncooked rice per serving; wash and drain rice (see Step 1 on page 22). Then cover rice with cold water (1½ cups or 350 ml water for first cup or 250 ml of rice, 1 cup or 250 ml for each additional cup or 250 ml) and soak 1–12 hours. Drain rice and spread it in a bamboo or metal steamer lined with cheesecloth. Steam over boiling water until rice is tender (about 25 minutes). Serve hot or at room temperature.

Dolmas

Rice-stuffed grape leaves are a blend of traditional Mediterranean foods.

1 jar (1 lb)	grape leaves, in brine	1 jar (450 g)
1/4 cup	olive oil	60 ml
1 tbl	butter	1 tbl
2 tbl each	minced shallot and garlic	2 tbl each
3 cups	cooked rice	700 ml
2 tbl each	dried currants and golden raisins	2 tbl each
1/4 cup each	chopped fresh mint and minced parsley	60 ml each
1 tsp	chopped fresh dill	1 tsp
2 oz	crumbled feta cheese	60 g
to taste	salt, pepper, and lemon juice	to taste
3–4 cups	hot chicken stock	700–900 ml
as needed	lemon wedges	as needed

1. Blanch grape leaves to remove excess brine; drain and refresh under cold running water. Drain well and pat dry.
2. In a small skillet over moderate heat, combine oil and butter. Add shallot and garlic and sauté until soft and slightly colored. Transfer to a large bowl and add rice, currants, raisins, mint, parsley, dill, and feta. Add salt, pepper, and lemon juice to taste. Toss well with a fork to blend.
3. Preheat oven to 350°F (175°C). Lay a grape leaf out flat; put about 1½ tablespoons filling near the base of leaf. Roll leaf, tucking in the sides as you roll. Repeat with remaining leaves. Transfer dolmas to a roasting pan large enough to hold them snugly. Cover with stock and bake, covered, for 20 minutes. Remove from oven and cool in stock. To serve, mound on a platter and accompany with lemon wedges.

Makes 15 to 20 dolmas.

Suppli al Telefono

At first bite the melted cheese in these rice croquettes (suppli) stretches into thin threads—like telephone lines (al telefono).

2 cups	cold cooked Arborio rice	500 ml
½ cup	grated Parmesan cheese	125 ml
3	eggs, lightly beaten	3
1 tsp	salt	1 tsp
⅛ tsp	freshly ground black pepper	⅛ tsp
¼ tsp	ground nutmeg	¼ tsp
1 tbl	finely chopped parsley	1 tbl
2 oz	prosciutto, finely chopped	60 g
as needed	flour	as needed
½ cup	dried bread crumbs	125 ml
2 oz	mozzarella cheese, cut into small dice	60 g
as needed	peanut oil, for deep-frying	as needed

1. In a large bowl combine rice, Parmesan cheese, 1 of the eggs, salt, pepper, nutmeg, parsley, and prosciutto. In a small bowl beat remaining 2 eggs with a tablespoon of water. Place ½ cup (125 ml) flour and the bread crumbs in 2 separate shallow dishes.

2. Flour hands and take a heaping tablespoon of rice mixture in palm of hand. Place about a teaspoon of diced mozzarella in center of rice. Pack another heaping tablespoon of rice mixture over cheese and press into a compact ball about 1½ inches (3.75 cm) in diameter. Roll ball in flour and shake off excess. Dip ball in beaten egg, coat with bread crumbs, and set on plate. Repeat with remaining ingredients. Chill 2 hours.

3. Into a large saucepan pour oil to a depth of 3 inches (7.5 cm) and heat to 375°F (190°C). Deep-fry balls in batches until golden brown (about 3 minutes), turning occasionally. Remove and drain on paper towels. Serve warm.

Makes 12 balls.

Starters

A rice soup or salad is a fine prelude to the main course or, served with fresh bread and a simple dessert, a satisfying light meal. Look within this section for mealtime starters made with rice, from a rustic Italian *zuppa* to a composed salad with flavors from the south of France.

Breton Mussel and Shrimp Soup with Rice

This sensational French soup makes a fine first course.

2½ cups	dry white wine	600 ml
2	onions, thinly sliced	2
1 stalk	celery, coarsely chopped	1 stalk
1	bay leaf	1
6 sprigs	parsley	6 sprigs
1 tsp	salt	1 tsp
¼ tsp	black peppercorns	¼ tsp
1 tbl	lemon juice	1 tbl
1 lb	shrimp, in shells	450 g
3 lb	mussels in shells, scrubbed	1.4 kg
1½–2 cups	fish or chicken stock	350–500 ml
⅛ tsp	saffron threads	⅛ tsp
4 tbl	butter	4 tbl
2	shallots, finely chopped	2
½ lb	mushrooms, thinly sliced	225 g
1 clove	garlic, minced	1 clove
⅛ tsp	cayenne pepper	⅛ tsp
1 can (6 oz)	tomato paste	1 can (175 ml)
¼ cup	chopped parsley	60 ml
3 cups	hot cooked long-grain rice	700 ml

1. In a large saucepan, combine 1½ cups (350 ml) water with ½ cup (125 ml) of the wine, 1 sliced onion, celery, bay leaf, 3 parsley sprigs, salt, black peppercorns, and lemon juice. Bring to a boil, cover, reduce heat, and simmer 5 minutes. Add shrimp. When mixture returns to a boil, remove pan from heat, cover, and let stand 10 minutes. Drain shrimp, reserving cooking liquid. Peel and devein shrimp; set aside.

2. In a kettle combine mussels, remaining parsley sprigs, and 1 cup (250 ml) of the wine. Bring to a boil over medium heat, cover, reduce heat, and simmer until mussels open (6–8 minutes). Discard any mussels that remain closed.

3. Remove mussels from liquid, reserving liquid and a few mussels in shells for garnish. Remove remaining mussels from shells. Pinch out and discard the beard from any mussel that has one. Add mussels to shrimp; set aside.
4. Strain mussel and shrimp cooking liquids through cheesecloth. Measure the combined liquid and add stock to make 6 cups (1.4 l). Place saffron in a small bowl and add ¼ cup (60 ml) of the hot shellfish liquid. Set aside to steep.
5. In a saucepan over medium heat, melt butter. Add shallots, mushrooms, and the other sliced onion; cook, stirring often, until soft and lightly browned. Mix in garlic, cayenne, and tomato paste. Add the shellfish liquid, saffron mixture, and remaining 1 cup (250 ml) wine.
6. Bring slowly just to a boil, stirring occasionally, until soup is hot. Mix in cooked mussels and shrimp. Taste and add salt if needed. Mix in chopped parsley. Ladle soup into warm bowls of hot rice. Garnish with mussels in shells.

Serves 6.

Chicken-Rice Soup

Make this flavorful soup (see photo on page 7) in 15 minutes.

4 cups	chicken stock	900 ml
6	canned water chestnuts, thinly sliced	6
2	green onions, thinly sliced	2
1 cup	cooked white rice	250 ml
to taste	salt and white pepper	to taste

In a medium saucepan over medium heat, bring stock to a gentle boil. Add water chestnuts and green onions. Reduce heat to medium-low and simmer 7 minutes. Add rice and heat through. Season with salt and pepper to taste, and serve.

Serves 4.

Sizzling Rice Soup

Refrying the Sizzling Rice Cakes (see page 16) just before serving makes this chicken-vegetable soup snap, crackle, and pop.

6 cups	chicken stock	1.4 l
1 stalk	celery, thinly sliced on the diagonal	1 stalk
1	onion, thinly sliced	1
¼ cup	sliced water chestnuts	60 ml
1 cup	sliced mushrooms	250 ml
1 cup	snow peas	250 ml
2 cups	diced cooked chicken	500 ml
1 tbl each	soy sauce and dry sherry	1 tbl each
1 tsp	Asian sesame oil	1 tsp
to taste	salt	to taste
as needed	peanut oil, for frying	as needed
1 recipe	Sizzling Rice Cakes (see page 16)	1 recipe

1. In a kettle heat chicken stock to boiling. Add celery, onion, water chestnuts, and mushrooms. Cover, reduce heat to medium, and cook for 5 minutes.
2. Add snow peas and chicken to soup; cook, uncovered, until chicken is heated through and snow peas turn bright green (3–5 minutes). Blend in soy sauce, sherry, and sesame oil. Add salt to taste.
3. Ladle soup into warm bowls; place a hot Sizzling Rice Cake in each bowl and serve at once.

Serves 6.

Avgolemono

This streamlined version of a classic Greek soup gets almost any meal off to a delicious start. Tart and refreshing, this soup takes just 15 minutes to prepare in the microwave.

3 cups	chicken stock	700 ml
1 cup	cooked rice	250 ml
2 tbl	butter	2 tbl
3	egg yolks	3
3 tbl	lemon juice	3 tbl
as needed	chopped parsley, for garnish	as needed

1. In a 3-quart (2.7-l) microwave-safe lidded dish, combine 2 cups of the chicken stock, rice, and butter. Cover and microwave on full power until stock is hot (5–6 minutes).

2. In a 2-cup (500-ml) microwave-safe container, microwave the remaining 1 cup (250 ml) stock on full power until boiling (2–3 minutes). In a small bowl beat egg yolks until light and fluffy. Gradually beat in lemon juice and the 1 cup (250 ml) hot stock.

3. Gently stir egg yolk mixture into cooked rice. Cover soup and let stand 5 minutes. Ladle into warm bowls and garnish with parsley. Serve at once.

Serves 6.

Seafood Filé Gumbo

Cooked white rice is the classic complement to gumbo (see top of photo on page 12).

1 cup each	vegetable oil and flour	250 ml each
4 cups	chopped onion	900 ml
1 cup	chopped celery	250 ml
3½ cups	chopped green and red bell pepper	800 ml
2 cups	chopped green onions	500 ml
3 tbl	minced garlic	3 tbl
½ cup	chopped parsley	125 ml
6	bay leaves	6
1 tsp	thyme	1 tsp
½ tsp each	oregano and sage	½ tsp each
½ tsp	cayenne pepper	½ tsp
12 cups	chicken or fish stock	2.7 l
¼ cup each	tomato paste and lemon juice	60 ml each
1½ lb	thinly sliced andouille sausage	680 g
¾ lb	medium shrimp, peeled and deveined	350 g
4–5	small blue crabs, broken in half	4–5
24	oysters, shucked	24
to taste	salt and freshly ground black pepper	to taste
2–3 tbl	filé powder	2–3 tbl

1. In a Dutch oven over medium heat, cook oil and flour until lightly browned. Add onion and celery and brown mixture for 10 minutes, stirring frequently. Add bell peppers, green onions, garlic, and parsley; cook another 5 minutes. Add bay leaves, thyme, oregano, sage, and cayenne. Cook another 5 minutes. Add stock, tomato paste, and lemon juice. Bring soup to a boil. Reduce heat and simmer 1 hour.

2. Add andouille and cook for 10 minutes. Add shrimp and crabs and simmer 5 minutes. Add oysters and cook 4–5 minutes. Add salt and pepper to taste. Remove gumbo from heat and mix in filé powder just before serving.

Serves 8.

Zuppa di Riso

This light Bolognese soup of vegetables and rice could precede a split grilled chicken or a hearty salad.

2 bunches	green or red chard	2 bunches
1/4 cup	olive oil	60 ml
1/4 cup	minced onion	60 ml
3 cups	chicken stock	700 ml
1/2 cup	Arborio rice, washed and drained	125 ml
1/3 cup	loosely packed fresh basil leaves	85 ml
as needed	freshly grated Parmesan cheese	as needed

1. Wash chard well. Remove stems, cut into 1/4-inch (.6-cm) pieces and set aside. Bring a large pot of salted water to a boil and blanch the chard leaves for 10–15 seconds. Cool in ice water, then drain, squeeze dry, and chop coarsely. Blanch the stems in the same boiling water, cool them in ice water, then drain and dry.

2. In a stockpot over moderate heat combine oil with onion; sauté until translucent and soft (about 3–5 minutes). Add chard stems and stir to coat with oil. Add chicken stock and rice and bring to a simmer. Cover and simmer gently until rice is just tender (about 15 minutes). Add chopped chard and heat through gently.

3. To serve, roll the basil and cut into fine strips. Ladle soup into warm bowls and garnish with basil and Parmesan cheese.

Serves 6.

Rice and Sprouts Salad

A cool, crunchy rice salad makes a fine first course or light meal on a warm day.

1/3 cup	rice wine vinegar	85 ml
2 tsp	sugar	2 tsp
1/2 tsp each	dried tarragon and salt	1/2 tsp each
pinch	white pepper	pinch
1/3 cup	olive oil	85 ml
1	red onion, slivered	1
2 cups	mung bean sprouts	500 ml
1/4 tsp each	salt and Asian sesame oil	1/4 tsp each
1/2 cup	long-grain rice, washed and drained	125 ml
1/4 cup	golden raisins	60 ml
1/2 cup	toasted slivered almonds	125 ml
2	green onions, thinly sliced	2
1/2 cup	chopped red or green bell pepper	125 ml
as needed	lettuce, for lining bowl	as needed

1. In a medium bowl combine vinegar, sugar, tarragon, salt, and pepper; stir until sugar dissolves. Using a whisk or fork, gradually beat in olive oil until well combined. Set dressing aside.

2. In a bowl mix red onion, bean sprouts, and dressing. Cover and refrigerate 1–2 hours.

3. In a medium saucepan over high heat, bring 1 cup (250 ml) water to a boil. Add salt, sesame oil, rice, and raisins. Cover, reduce heat, and simmer 20–25 minutes, or until rice is just tender. Let cool to room temperature.

4. Add rice mixture to bean sprout mixture along with almonds, green onions, and bell pepper. Mix lightly. Cover and refrigerate for at least 2 hours. Serve in a lettuce-lined bowl.

Serves 4.

Four-Seasons Rice Salad

You can put rice salad on the menu again and again without having too much of a good thing: Just vary the recipe for your favorite basic rice salad by accenting it with whatever fresh produce is in season. In spring try a mixture of asparagus, snow peas, and chopped green onions. Cucumbers, bell peppers, and sweet corn make a fine summertime combination. Autumn offers a colorful blend of carrots, broccoli, and almonds or, for a sweet taste, apples, dates, and cranberries. Perk up winter rice salads with spinach leaves and celery tossed with pomegranate seeds, tangerines, or oranges.

Lentil-Pilaf Salad

For this satisfying pilaf salad the rice is first sautéed, then baked in a seasoned liquid.

2½ cups	chicken stock	600 ml
½ cup	raisins	125 ml
1½ tbl	butter	1½ tbl
1½ tbl	olive oil	1½ tbl
½	onion, diced	½
¾ cup	long-grain rice, washed and drained	175 ml
½ cup	dried lentils	125 ml
⅛ tsp each	ground cardamom, ground cinnamon, ground cloves, ground coriander, ground cumin, and freshly ground black pepper	⅛ tsp each
¼ lb	fresh green beans, trimmed and cut into 1-inch (2.5-cm) lengths	115 g
½ cup	frozen tiny peas	125 ml
1	carrot, peeled and diced	1
½ cup	diced red or green bell pepper	125 ml
as needed	chopped cilantro (coriander leaves), as garnish	as needed

1. Preheat oven to 375°F (190°C). In a large saucepan over medium heat, combine chicken stock and raisins and bring to a boil. Remove from heat and keep warm.

2. In a large Dutch oven over medium-high heat, melt butter with oil. Add onion and sauté until lightly browned. Stir in rice, lentils, cardamom, cinnamon, cloves, coriander, cumin, and pepper. Stir and cook 2–3 minutes. Add green beans, peas, carrot, and bell pepper; pour in stock and raisins. Cover and bake until all liquid has been absorbed, about 20 minutes. Chill and serve garnished with cilantro.

Serves 6.

Rice Salad Niçoise

Ingredients from the Côte d'Azur make a refreshing salad.

2 tbl	red wine vinegar	2 tbl
1½ tsp	lemon juice	1½ tsp
2 tsp	Dijon mustard	2 tsp
1 clove	garlic, minced	1 clove
⅓ cup	olive oil	85 ml
1 tbl	chopped parsley	1 tbl
to taste	salt and freshly ground black pepper	to taste
2½ cups	cooked long-grain rice	600 ml
1 jar (6 oz)	marinated artichoke hearts	1 jar (175 ml)
1 can (2 oz)	anchovy fillets, drained	1 can (60 ml)
1	green bell pepper, chopped	1
1 stalk	celery, finely chopped	1 stalk
4	green onions, thinly sliced	4
¼ cup	chopped Niçoise olives	60 ml
½ cup	chopped, peeled cucumber	125 ml
1	tomato, seeded and chopped	1
1 can (6½ oz)	chunk light tuna	1 can (200 ml)
2	hard-cooked eggs, sliced	2
as needed	lettuce leaves and tomato wedges	as needed

1. In a medium bowl mix vinegar, lemon juice, mustard, garlic. Whisk in oil until mixture is slightly thickened. Stir in parsley and add salt and pepper to taste. Set aside.

2. In a large bowl, lightly mix rice with artichoke hearts and their marinade. Cover and refrigerate until cool. Reserving 3 anchovy fillets for garnish, chop remaining anchovies and add to rice with remaining ingredients except lettuce and garnishes. Add dressing and mix lightly. Cover and refrigerate 2 hours. Serve in a lettuce-lined bowl. Garnish with tomato wedges and reserved anchovies.

Serves 6.

Off-the-Shelf Rice Salad

The next time you go food shopping, stock up on staples for satisfying rice salads you can make entirely from items in the cupboard. Besides long-grain rice (the best for salads), lay in a supply of ingredients for dressings: vinegars (rice, balsamic, wine), oils (olive, Asian sesame, walnut), and soy sauce, garlic, and dried herbs. Add an assortment of good-quality canned and bottled foods—tuna or crab, pickles, marinated artichoke hearts, roasted peppers, beans (garbanzos, kidney beans, and flageolets), and perhaps nuts (peanuts, cashews, and pecans)—and you have the makings for a dozen different variations.

Main Courses

Rice makes imaginative, economical, healthful main-course fare, as the recipes in this section demonstrate. From Europe, the Far East, and the United States come meal-in-a-pot favorites like spicy paella, risotto, stir-fried specialties, and other rice entrées to highlight any menu.

NEW ORLEANS RED BEANS AND RICE

This is the traditional Monday-night dinner of New Orleans.

1 lb	dry red beans	450 g
2	meaty ham hocks	2
8 cups	beef or chicken stock	1.8 l
4	bay leaves	4
½ tsp	thyme	½ tsp
1 tsp	cayenne pepper	1 tsp
as needed	freshly ground black pepper	as needed
1½ lb	sliced andouille sausage	680 g
2 cups	chopped onion	500 ml
½ cup	chopped celery	125 ml
1	green bell pepper, chopped	1
1 bunch	green onions, chopped	1 bunch
1 tbl	minced garlic	1 tbl
to taste	salt	to taste
to taste	red wine vinegar	to taste
4 cups	cooked hot rice	900 ml

1. Wash beans and soak overnight in water to cover. The next day drain beans and rinse. Place beans, ham hocks, and stock in a heavy stockpot. Bring to a boil and skim any scum that collects on the surface. Reduce heat to a simmer and add bay leaves, thyme, cayenne, and 2 teaspoons black pepper. Simmer for 30 minutes.

2. In a large skillet on high heat, sauté sausage for 5 minutes. Add chopped onion, celery, bell pepper, green onions, and garlic. Cook for 15 minutes, then add to red beans. Continue to cook beans until they soften (about 1 hour more). Let cool and refrigerate, covered, overnight or for up to 4 days.

3. When ready to serve, bring beans to a simmer. Season with salt, pepper, and vinegar to taste. Place about ½ cup (125 ml) rice on each plate and spoon beans and their juice over rice.

Serves 8.

Hoppin' John

In many areas of the Deep South this dish is served with greens and corn bread on New Year's Day to ensure good luck.

2 cans (15 oz each)	black-eyed peas, undrained	2 cans (450 ml each)
6 strips	bacon, diced	6 strips
1	onion, chopped	1
as needed	salt	as needed
¾ cup	long-grain white rice, washed and drained	175 ml
to taste	freshly ground black pepper	to taste
to taste	Louisiana-style hot sauce	to taste
½ cup	minced green onions, for garnish	125 ml
3 tbl	minced parsley, for garnish	3 tbl

1. Place peas and their liquid in a large pot. In a separate pan, sauté bacon until crisp; crumble bacon and add it to peas, reserving rendered drippings. Add onion, ½ teaspoon salt, and ½ cup (125 ml) water. Bring to a boil, lower heat, and simmer until onion is tender (about 8 minutes).

2. In separate pot, cover rice with cold water. Bring to a boil, stir once, cover, and reduce heat. Simmer for 20 minutes.

3. Add cooked rice to peas. Stir in 2 tablespoons reserved bacon fat; add salt, pepper, and hot sauce to taste. Cover and simmer about 15 minutes longer. To serve, garnish with green onions and parsley.

Serves 6.

Bon Temps Jambalaya

Keep the good times rolling with this Louisiana classic.

2	bay leaves	2
1½ tsp each	salt, ground red pepper, dried oregano, and white pepper	1½ tsp each
¾ tsp	dried thyme	¾ tsp
2½ tbl	olive oil	2½ tbl
⅔ cup	chopped smoked ham	150 ml
½ cup	chopped andouille sausage	125 ml
1½ cups	chopped onion	350 ml
1 cup	chopped celery	250 ml
¾ cup	chopped green bell pepper	175 ml
1½ tsp	minced garlic	1½ tsp
4 cups	peeled, chopped tomatoes	900 ml
¾ cup	tomato sauce	175 ml
2 cups	fish or chicken stock	500 ml
½ cup	chopped green onions	125 ml
2 cups	uncooked rice, washed and drained	500 ml
1 lb	firm-fleshed fish fillets, cut into bite-sized pieces	450 g
18	medium oysters	18
18	medium shrimp, peeled and deveined	18

1. Preheat oven to 350°F (175°C). In a small bowl combine bay leaves, salt, red pepper, oregano, white pepper, and thyme. Set mixture aside.

2. Into a large saucepan over medium heat, pour oil. Add ham and sausage; sauté until crisp (5–8 minutes), stirring frequently. Add onion, celery, and bell pepper; sauté until tender but still firm (about 5 minutes), stirring occasionally and scraping pan bottom well.

3. Add seasoning mixture and garlic; cook about 3 minutes, stirring constantly and scraping pan bottom as needed. Add tomatoes and cook about 7 minutes, stirring frequently. Add tomato sauce; cook about 7 minutes more, stirring often.

4. Stir in stock and bring to a boil. Then stir in green onions and cook about 2 minutes, stirring briefly. Add rice, fish, oysters, and shrimp; stir well and remove from heat.
5. Transfer mixture to an ungreased 9- by 13-inch (22.5- by 32.5-cm) baking pan. Cover pan snugly with aluminum foil and bake until rice is tender but not mushy (20–30 minutes). Remove from oven. Let pan sit a few minutes, still covered, to allow rice to absorb any remaining liquid. Remove bay leaves and spoon jambalaya onto serving plates.

Serves 6.

Brown Jambalaya

The rich color of this tomatoless jambalaya comes from the browned onions and smoked meats (see bottom of photo on page 12).

2 lb	thinly sliced andouille sausage	900 g
½ lb	diced tasso or smoked ham	225 g
4 cups	chopped onion	900 ml
½ cup	chopped celery	125 ml
2 tbl	soy sauce	2 tbl
1 cup	chopped green onions	250 ml
1 tbl	minced garlic	1 tbl
1 cup	finely chopped green bell pepper	250 ml
½ tsp	cayenne	½ tsp
2	bay leaves	2
½ tsp each	thyme and sage	½ tsp each
3 cups	chicken stock	700 ml
2 cups	long-grain or converted rice, washed and drained	500 ml
3 cups	diced cooked chicken	700 ml
to taste	salt and freshly ground black pepper	to taste

1. In a large saucepan over medium heat, sauté andouille and tasso until lightly browned (about 10 minutes). Remove meats from pan and set aside.

2. Add onion and celery to the pan and sauté over medium heat, stirring occasionally, until deep brown (about 30 minutes). Add soy sauce, green onions, garlic, and bell pepper and sauté for 5 minutes. Add cayenne, bay leaves, thyme, sage, and reserved meats. Transfer mixture to large bowl and set aside.

3. Add stock to the pan and bring to a boil; add rice. Reduce heat to a simmer, cover, and cook about 20 minutes. Stir in chicken and meat mixture; cook until almost all the liquid is absorbed and the rice is cooked but not mushy (about 10 minutes). Add salt and pepper to taste.

Serves 8.

Minnesota Pilaf with Cashew Gravy

Wild rice makes an aromatic Midwestern pilaf.

1 cup	minced onion	250 ml
4 tsp	olive oil	4 tsp
¼ cup	dry sherry or white wine	60 ml
½ cup each	diced carrots and celery	125 ml each
1 tsp	minced garlic	1 tsp
¼ cup	pine nuts	60 ml
2 cups	wild rice, washed and drained	500 ml
1 cup	long-grain brown rice, washed and drained	250 ml
3 cups	vegetable or chicken stock	700 ml
¼ tsp each	thyme and sage	¼ tsp each
1 tsp	soy sauce or tamari	1 tsp
1 cup	toasted whole cashews	250 ml
2 tsp	whole wheat pastry flour	2 tsp
2 tsp	grated ginger	2 tsp
1 tsp	salt	1 tsp
1 tbl	minced parsley	1 tbl
1 tbl	miso	1 tbl

1. In a large saucepan over medium-high heat, sauté onion in 2 teaspoons oil and sherry for 5 minutes, then add carrots, celery, and garlic. Cover and steam for 2 minutes. Add pine nuts and rices, and stir-fry for 1 minute. Pour in stock and bring to a boil. Lower heat to medium and simmer, uncovered, for 15 minutes, then cover and steam 20 minutes. Add thyme, sage, and soy sauce. Set aside.

2. In a blender purée cashews in 2 cups (500 ml) water. Set aside. In a saucepan over medium heat, cook remaining oil and flour, stirring frequently, for 2 minutes. Add cashew mixture, ginger, salt, and parsley. Cook until thick, whisking frequently. Remove some of the cashew gravy to a small bowl and mix with miso until smooth. Stir miso mixture into gravy and serve over pilaf.

Serves 6.

Shrimp Okra Pilau

Pilau is the mid-South's version of Spanish paella. Properly cooked, okra develops a sensuous texture and distinctive flavor that perfectly complements the other vegetables in this rice dish.

3 strips	bacon	3 strips
1	onion, finely chopped	1
1	green bell pepper, finely chopped	1
1 lb	fresh okra, trimmed and sliced into rounds, or 1 package (12 oz/350 g) sliced frozen okra	450 g
1 cup	long-grain white rice, washed and drained	250 ml
1 lb	fresh tomatoes, peeled and coarsely chopped, or 1 can (14 oz/425 ml) tomatoes, coarsely chopped	450 g
½ tsp	salt and freshly ground black pepper	½ tsp
1 lb	peeled medium shrimp	450 g
as needed	cayenne pepper	as needed

1. Cook bacon in a large skillet until crisp. Drain on paper towels, crumble, and reserve.
2. In bacon fat remaining in the skillet (about 3 tablespoons), sauté onion and bell pepper over high heat, stirring until wilted (about 5 minutes). Add okra and cook over low heat for 10 minutes, stirring frequently.
3. Place rice in saucepan with cover and add 2 cups (500 ml) cold water. Bring to a boil, stir briefly, and cook over low heat for 12 minutes. Drain rice and add to okra along with tomatoes, ½ teaspoon each salt and pepper, and bacon. Cover skillet and continue to cook over low heat for 8 minutes. Add shrimp and cayenne and cook 7 minutes more. Serve hot.

Serves 4.

Roast Chicken with Pilaf Stuffing

The rice stuffing for this chicken is equally good served as a side dish. Simply spoon the pilaf into a lidded casserole and bake at 350°F (175°C) for half an hour.

¼ cup	olive oil	60 ml
1	onion, finely chopped	1
1 cup	long-grain rice, washed and drained	250 ml
½ cup	fresh orange juice	125 ml
to taste	salt and freshly ground black pepper	to taste
1	apple, finely chopped	1
½ cup	raisins	125 ml
¼ tsp	ground cinnamon	¼ tsp
½ cup	slivered almonds, toasted	125 ml
one (4 lb)	roasting chicken	one (1.8 kg)

1. In a frying pan over low heat, heat 3 tablespoons of the oil. Add onion and cook, stirring often, until tender (about 5 minutes). Raise heat to medium, add rice, and sauté for 2 minutes. Add 1½ cups (350 ml) hot water, orange juice, ½ teaspoon each salt and pepper, and bring to a boil. Reduce heat to low, cover, and cook for 10 minutes.

2. Add apple and raisins to rice and stir very lightly with a fork. Cover and cook for 5 minutes. Stir in cinnamon and almonds. Add more salt and pepper if needed.

3. Preheat oven to 425°F (220°C). Sprinkle chicken with salt and pepper. Spoon enough rice stuffing into chicken to fill it, but do not pack it too tightly.

4. Set chicken on a rack in a roasting pan and roast until juices run clear when a skewer is inserted into thickest part of leg (about 1 hour). Let stand 5–10 minutes before serving.

Serves 4.

Rice Venetian

This classic scampi dish makes an exquisite but satisfying supper.

1 tbl	olive oil	1 tbl
2 tbl	butter	2 tbl
2 tbl	minced shallot	2 tbl
1 cup	Arborio rice, washed and drained	250 ml
2 cups	fish or chicken stock	500 ml
1/4 cup	tomato purée	60 ml
1 1/4 lb	medium shrimp, peeled and deveined	570 g
to taste	salt and freshly ground black pepper	to taste
2 tbl	minced parsley	2 tbl

1. In a saucepan over moderately low heat, combine olive oil and butter. Add shallot and sauté gently until softened (about 5 minutes). Add rice and cook, stirring, for 1 minute. Add fish stock and tomato purée. Bring to a boil. Cover, reduce heat to low, and cook 8 minutes.

2. Arrange shrimp on top of rice, cover, and cook until shrimp are just done (about 7 minutes). Add salt and pepper to taste. Transfer to a warm serving platter and garnish with parsley.

Serves 4.

Rice for Risotto

From the Po River region in northern Italy comes Arborio, the variety of rice used to make the flavorful, slow-cooked Italian specialty known as risotto. Arborio has the unique capacity to absorb as much as five times its weight in liquid while maintaining its shape during the cooking process. It is this quality that produces the enticing combination of creamy and chewy textures in an authentic risotto *(See Preparing Risotto on page 85).*

Risotto with Smoked Salmon

Traditional risotto demands Arborio rice (see page 52) and a special cooking method (see page 85). Properly cooked, the center of each grain is al dente (offering a slight resistance to the tooth), while the outside is delectably creamy. For variations, use cooked clams, mussels, or calamari instead of salmon.

9 tbl	unsalted butter	9 tbl
1/4 cup	minced onion	60 ml
2 cups	Arborio rice, washed and drained	500 ml
3–5 cups	fish or chicken stock	700–1.1 l
2/3 cup	slivered smoked salmon	150 ml
4 tbl	freshly grated Parmesan cheese	4 tbl
to taste	ground white pepper	to taste
3 tbl	chopped parsley	3 tbl

1. In a large saucepan over medium heat, melt 6 tablespoons of the butter, add onion, and cook slowly until onion is soft. Add rice to pan, stirring until grains of rice are well coated.

2. In a small saucepan, heat the stock. When it is hot, add 1 cup (250 ml) of the stock to the butter and rice mixture and stir constantly until liquid is absorbed. Add stock, 1/2 cup (125 ml) at a time, stirring continuously and allowing rice to absorb liquid before adding more. Test rice for doneness as it cooks. When properly cooked, the mixture is creamy, but each grain of rice retains its shape and is al dente in the center.

3. Add salmon and the remaining butter. Stir well, sprinkle with Parmesan cheese, pepper to taste, and parsley. Serve at once.

Serves 8.

Basque Paella

A paellero is the traditional lidded dish in which paella cooks.

1 tbl	olive oil	1 tbl
1 cup	boneless chicken, diced	250 ml
1 cup	basmati rice, washed and drained	250 ml
½ cup	chopped onion	125 ml
2 cloves	garlic, minced	2 cloves
1 cup	sliced bell pepper	250 ml
½ cup	diced tomato	125 ml
1	red snapper fillet, cut into 1-inch (2.5-cm) pieces	1
2 cups	hot chicken stock	500 ml
1 tsp each	salt and saffron threads	1 tsp each
¾ tsp	dried oregano	¾ tsp
6	large prawns, peeled and deveined	6
1 cup	peas, fresh shelled or frozen	250 ml
6	artichoke hearts, unmarinated	6
6	clams in their shells, scrubbed	6

1. In large skillet over medium-high heat, heat oil and cook chicken until just opaque (about 6–8 minutes). Remove chicken to platter. Reserve skillet.

2. Soak rice in 2 cups (500 ml) boiling water for 10 minutes, then drain, reserving soaking water. In reserved skillet sauté onion over medium heat until soft (about 5 minutes). Add garlic, bell pepper, and tomato; sauté for 5 minutes. Add snapper, soaked rice with water, stock, salt, saffron, and oregano. Bring to a boil, then lower heat to medium. Cover and simmer until rice is tender (about 15 minutes).

3. Arrange remaining ingredients on top of rice. Cover and cook over medium heat until prawns turn bright pink and clam shells open (about 10 minutes). Discard any unopened clams. Serve hot.

Serves 6.

Szechuan Spareribs with Toasted Rice

A toasted rice coating for marinated spareribs is a Szechuan specialty. For a colorful presentation, arrange the ribs on half a steamed winter squash.

1½ lb	pork spareribs, chopped into 1-inch (2.5-cm) strips	680 g
1 tsp	minced fresh ginger	1 tsp
1 tsp	minced garlic	1 tsp
¼ tsp	salt	¼ tsp
1 tsp	sugar	1 tsp
2 tbl	soy sauce	2 tbl
1 tbl	rice wine or dry vermouth	1 tbl
2 tsp	hot bean sauce	2 tsp
½ tsp	five-spice powder	½ tsp
1 tsp	Asian sesame oil	1 tsp
3 tbl	glutinous rice, washed and drained	3 tbl
as needed	cooked white long-grain rice, for accompaniment	as needed

1. Cut between ribs to form 1-inch (2.5-cm) squares of meat, each with some bone. Place in a large bowl and add ginger, garlic, salt, sugar, soy sauce, wine, bean sauce, five-spice powder, and sesame oil. Mix well; set aside.

2. In an ungreased wok over low heat, toast glutinous rice until lightly browned (6–8 minutes), stirring occasionally. Remove from heat and let cool. Transfer glutinous rice to a mortar or food processor and grind to consistency of cornmeal. Drain ribs. Pour rice crumbs over ribs and toss lightly to coat.

3. Prepare wok for steaming. Arrange no more than 2 layers of coated ribs on a shallow heat-resistant plate. Cover and steam over medium-high heat 45 minutes. Serve hot with cooked rice.

Serves 4.

Chicken and Rice in a Clay Pot

Cook this dish in any flameproof casserole.

2	shallots, minced	2
2 tbl	minced garlic	2 tbl
1 tbl	minced ginger	1 tbl
2 lb	chicken parts	900 g
2 tbl	soy sauce	2 tbl
1–2 tbl	Asian sesame oil	1–2 tbl
to taste	freshly ground black pepper	to taste
3 cups	chicken stock	700 ml
8	dried shiitake mushrooms, soaked and drained, liquid reserved	8
2 cups	long-grain rice, washed and drained	500 ml

1. Combine shallots, garlic, and ginger and mince together finely. Place chicken in bowl. Add shallot mixture and soy sauce, toss to coat chicken pieces, and marinate 30 minutes.

2. Pour oil into a wok over medium heat. Scrape excess marinade from chicken pieces and cook a few at a time until lightly browned on all sides. Remove chicken from wok. Add to wok the reserved marinade, ½ teaspoon pepper, stock, and reserved mushroom liquid. Bring to a boil, scraping up any bits of chicken clinging to wok.

3. Arrange chicken pieces in braising pot and pour in boiling liquid. Cover and simmer until chicken is quite tender (about 20 minutes). Remove chicken from pot and allow to cool. When cool enough to handle, shred meat from bones.

4. Skim fat from liquid and add rice. Halve mushrooms and add to rice mixture. Bring to a boil, reduce heat, and simmer, stirring occasionally, until liquid is nearly absorbed. Adjust seasoning if necessary. Add chicken, cover, and cook over low heat 10 minutes more. Turn off heat; let stand 15 minutes before serving.

Serves 4.

Riz à la Basquaise

The Basque region, with its rustic cuisine and Spanish influences, is one of the few areas in France where rice dishes are common.

¼ lb	diced bacon	115 g
2 cups	coarsely chopped red and green bell pepper	500 ml
2 tbl	olive oil	2 tbl
½ lb	spicy sausage links	225 g
1½ cups	minced onion	350 ml
1	jalapeño chile, seeded and minced	1
1½ cups	long-grain white rice, washed and drained	350 ml
3 cups	hot chicken stock	700 ml
2 cups	seeded, diced tomato	500 ml
3 tbl	minced parsley	3 tbl
to taste	salt and freshly ground black pepper	to taste

1. In a large skillet over moderately low heat, cook bacon until most of the fat is rendered. Add bell peppers, cover, and cook for 10 minutes.

2. In another large skillet over moderate heat, heat olive oil. Add sausage and brown lightly on all sides. Add onion and jalapeño, and reduce heat to low. Sauté for 10 minutes. Add rice and stir to coat with oil. Add stock and bring to a boil. Cover tightly and cook over low heat for 18 minutes.

3. Remove sausage from rice, slice into rounds, and return to rice. Drain excess fat from bacon, then stir peppers and bacon into rice. Combine tomato and parsley. Add to rice and toss well with a fork. Season to taste with salt and pepper. Transfer to a warm serving bowl and serve.

Serves 6.

Shrimp and Pork Fried Rice

Chinese shrimp paste gives authentic flavor to this stir-fry.

3 tbl	peanut oil	3 tbl
½ tsp	salt	½ tsp
1 tsp	Chinese-style wet shrimp paste	1 tsp
3 cups	cooked white long-grain rice	700 ml
½ tsp	sugar	½ tsp
1½ tbl	soy sauce	1½ tbl
2 tsp	oyster sauce	2 tsp
2	eggs, plus 1 egg yolk	2
½ cup	cooked bay shrimp	125 ml
½ cup	diced Chinese barbecued pork	125 ml
½ cup	diced cooked chicken	125 ml
½ cup	blanched peas	125 ml
1 cup	finely shredded lettuce	250 ml
½ cup	chopped green onion	125 ml

1. Preheat wok over medium-high heat. Pour in oil, add salt and shrimp paste, and stir until oil is fragrant (about 5 seconds). Add rice, tossing until grains are separate but not brown (about 3 minutes). Season with sugar, soy sauce, and oyster sauce; stir-fry to coat rice (about 1 minute).

2. Add eggs to wok; cook 1 minute, lightly stirring. Toss until flecks of egg appear throughout rice. Add remaining ingredients; toss until lettuce is wilted (about 2 minutes) and serve.

Serves 4.

Preparing "Fried Rice"

What started as a way to use leftover cooked rice has sparked a host of famous Asian and Asian-American dishes. "Fried rice" is a misnomer, since the rice is not actually fried but instead stir-fried. For best results, the cooked rice should be at least a couple of hours old—not freshly cooked.

Sausage and Fennel Fried Rice

Fresh fennel combines beautifully with sweet-spicy Cantonese-style sausage (lop cheong) and bok choy in this stir-fry. See Preparing "Fried Rice" on page 61.

2	Chinese sausages	2
¼ cup	peanut oil	60 ml
¼ cup	raw peanuts	60 ml
1 tsp each	minced ginger and garlic	1 tsp each
1 cup	sliced bok choy	250 ml
1	small fennel bulb, trimmed, cut crosswise into ¼-inch (.6-cm) slices	1
3 cups	cooked long-grain rice	700 ml
¼ cup	chicken stock	60 ml
to taste	salt	to taste

1. Slice sausages thinly on the diagonal. Steam on a rack or in a bowl for 10 minutes to render some of the fat. Discard fat.
2. Heat wok over medium heat; add oil. Fry peanuts until light brown. Remove and drain on paper towels.
3. Remove all but 2 tablespoons oil from pan. Add ginger and garlic; cook until fragrant. Add steamed sausage and bok choy; stir-fry 1 minute. Add fennel and cook 30 seconds more. Add rice and stir-fry, scraping pan thoroughly, until lightly browned. Add stock and ½ teaspoon salt; cook until nearly dry. Taste and add salt if needed; transfer to serving platter and scatter peanuts over top.

Serves 4.

Indonesian Fried Rice

This recipe makes a fine one-dish meal. The kecap manis (Indonesian soy sauce), dried shrimp, and fish sauce are available in Asian markets. See Preparing "Fried Rice" on page 61.

¼ lb	tender beef	115 g
1 tbl	kecap manis	1 tbl
4 tbl	peanut oil	4 tbl
2	eggs, lightly beaten	2
1 tbl	minced garlic	1 tbl
2	fresh red chiles, seeds and veins removed, diced	2
3 tbl	dried shrimp, soaked in water 15 minutes, drained, and minced	3 tbl
¼ lb	fresh shrimp, peeled and deveined	115 g
3 cups	cooked rice	700 ml
½ cup each	diced cabbage and tomato	125 ml each
1 tbl	fish sauce	1 tbl
as needed	sliced green onions, for garnish	as needed

1. Thinly slice beef across grain and cut into narrow 1½-inch (3.75-cm) strips. In a small bowl combine with kecap manis and marinate 15 minutes.

2. Heat a wok over medium heat and add 2 tablespoons of the oil. Add eggs and swirl pan to form a thin omelet. When eggs are almost set, remove the omelet to a plate and set aside. When cool cut into thin strips.

3. Add remaining oil to pan. Add garlic, chiles, and dried shrimp and stir-fry until fragrant. Add fresh shrimp and beef and stir-fry until shrimp begins to turn pink. Add rice to pan and stir. Turn heat to medium-high and stir-fry vigorously. When rice begins to brown, stir in cabbage and tomato and continue stir-frying until mixture is nearly dry. Sprinkle with fish sauce and transfer to serving platter. Garnish with green onions and egg strips.

Serves 4.

Mandarin Fried Rice

Colorful Mandarin-style fried rice, seasoned with a hint of ginger, can be prepared in minutes. See Preparing "Fried Rice" on page 61.

3	eggs	3
1¼ tsp	salt	1¼ tsp
4 tbl	peanut oil	4 tbl
1 tbl	freshly grated ginger	1 tbl
5 cups	cooked white long-grain rice	1.1 l
2	green onions, chopped	2
6 oz	cooked small bay shrimp	170 g
½ cup	cooked peas	125 ml

1. In a small bowl lightly beat eggs with ¼ teaspoon of the salt; set aside.

2. Preheat wok over medium-high heat. Pour in 2 tablespoons of the oil. When hot add reserved egg mixture. Stir-fry until egg is set but not dry (about 1 minute), tilting wok and pushing cooked egg up sides to allow uncooked egg to flow to the center. Remove from heat and transfer egg to a medium bowl. With a spatula chop egg into small pieces and set aside.

3. Replace wok over medium-high heat and add remaining 2 tablespoons oil, remaining 1 teaspoon salt, and ginger. Cook until oil is fragrant (about 30 seconds). Add rice and toss until grains are separate but not browned (2–3 minutes). Add remaining ingredients; toss together. Add reserved chopped egg and combine. Serve hot.

Serves 4.

Four-Treasure Rice

Serve this stir-fry with a salad for a light lunch. See Preparing "Fried Rice" on page 61.

6 tbl	peanut oil	6 tbl
1 tsp	salt	1 tsp
4 cups	cooked long-grain rice, cooled	900 ml
¾ cup	finely chopped onion	175 ml
1½ cups	chopped cooked shrimp	350 ml
1 cup	chopped crabmeat	250 ml
¼ cup	peas	60 ml
½ cup	coarsely chopped green onion	125 ml
3	eggs	3
¾ tsp	salt	¾ tsp

1. In a wok heat 4 tablespoons of the oil. Add the 1 teaspoon salt, then add rice, tossing until rice is well coated with oil. Stir-fry for 2 minutes. Add the ¾ cup (175 ml) chopped onion, shrimp, crab, peas, and the green onion. Toss and heat through.
2. Place a small pan over medium heat and add the remaining 2 tablespoons oil. Add eggs and the ¾ teaspoon salt and scramble lightly. Add scrambled eggs to rice and toss together. Adjust seasonings, if needed, and serve immediately.

Serves 4.

The Stir-Fry Pantry

Keeping a supply of stir-fry essentials on hand gives your repertoire of quick-meal ideas almost limitless potential. With some leftover rice, fresh snow peas or broccoli, fresh ginger, green onions, and an egg or two from the refrigerator; garlic, dried mushrooms, peanut or sesame oil, and soy sauce from the pantry; and chicken stock from the freezer or pantry, you can have an authentic stir-fry ready in minutes.

Kedgeree

This Anglo-Indian rice creation makes an irresistible brunch dish.

4 tbl	butter	4 tbl
1	onion, finely chopped	1
¾ tsp	Indian curry powder	¾ tsp
1 cup	long-grain rice, washed and drained	250 ml
½ tsp	salt	½ tsp
¾ lb	smoked salmon or haddock	350 g
2 tbl	flour	2 tbl
½ tsp	white pepper	½ tsp
1½ cups	chicken stock	350 ml
3	hard-cooked eggs, shredded	3
as needed	chopped parsley, for garnish	as needed

1. In a large, heavy frying pan over medium heat, melt 2 tablespoons of the butter and cook onion until soft but not brown. Stir in curry powder and rice until well combined. Sprinkle with salt. Add 2 cups (500 ml) water, cover, reduce heat, and simmer until rice is tender and liquid is absorbed (20–25 minutes).

2. Steam fish on a rack over gently boiling water until it separates easily into flakes (10–15 minutes). Flake fish, reserving a few large pieces for garnish; keep warm.

3. In a medium saucepan over moderate heat, melt remaining 2 tablespoons butter. Add flour and white pepper, stirring until bubbly. Remove from heat and gradually blend in chicken stock. Cook, stirring, until thickened and bubbling. Then boil gently, stirring occasionally, until reduced to about 1 cup (250 ml).

4. Mix flaked fish and sauce into the cooked rice. Spoon into a warm serving dish and top with reserved pieces of fish and shredded eggs. Sprinkle with parsley and serve at once.

Serves 4.

Accompaniments

Versatile side dishes featuring rice provide an appealing, nutritious way to round out any meal. Authentic Italian pilafs and risottos, a trio of Mexican "dry soups" made with rice, and other dishes in this section make what's served on the side every bit as special as the main course.

Rice Pilaf with Parmesan Cream

Pilafs are made of sautéed rice simmered in a flavorful cooking liquid. In this exciting dish, rice is sautéed with dried porcini mushrooms and pancetta (thick-sliced Italian bacon) and simmered with heavy cream.

1/4 cup	dried porcini mushrooms	60 ml
2 oz	pancetta, diced	60 g
3 tbl	unsalted butter	3 tbl
2	shallots, chopped	2
1/2 cup	fresh shelled peas, parboiled 2 minutes and drained	125 ml
1/2 cup	heavy cream	125 ml
1/2 cup	freshly grated Parmesan cheese	125 ml
pinch	freshly grated nutmeg	pinch
to taste	freshly ground black pepper	to taste
2 1/2 cups	cooked rice	600 ml

1. In a small bowl cover mushrooms with warm water and soak until soft (about 3 hours). Drain mushrooms, reserving 1/4 cup (60 ml) liquid; coarsely chop mushrooms and set aside.

2. Preheat wok over medium heat. Sauté pancetta until brown. Spoon off and discard all but 1 tablespoon fat. Add butter and heat until it begins to sizzle. Add shallots, reserved mushrooms, and peas; sauté 5 minutes.

3. Add reserved mushroom liquid, cream, cheese, nutmeg, and dash of pepper. Carefully stir in rice and simmer, stirring occasionally, until rice is warm (5–8 minutes). Add more pepper to taste and serve.

Serves 4.

Pilaf Nouvelle

This colorful microwaved pilaf is perfect for busy times when you need an easy side dish that doesn't require much attention. You can cook and serve this pilaf in the same dish, saving cleanup time.

1 cup	long-grain rice, washed and drained	250 ml
2 tbl	butter	2 tbl
1	carrot, chopped	1
½ cup	sliced green onion	125 ml
1 clove	garlic, minced	1 clove
1 cup	sliced mushrooms	250 ml
2 tsp	chicken bouillon	2 tsp
to taste	salt and freshly ground black pepper	to taste
as needed	chopped parsley, for garnish	as needed

1. In a 3-quart (2.7-l) microwave-safe casserole, combine rice, 2 cups (500 ml) hot water, and butter. Cover and microwave on full power until rice has boiled 5 minutes (12–15 minutes total).
2. Add carrot, green onion, garlic, mushrooms, and bouillon. Cover and microwave on full power until rice and vegetables are tender and water is absorbed (6–8 minutes).
3. Fluff rice with a fork. Season with salt and pepper to taste and garnish with chopped parsley.

Serves 4.

PILAF MINCEUR

Wild, basmati, and brown rice team up with aromatic vegetables and herbs in this lowfat pilaf. Using sherry instead of oil to sauté the vegetables lowers the fat content of this dish and further enhances the aroma and flavor. Serve with poached salmon or broiled chicken.

1 cup	chopped onion	250 ml
2 tsp	minced garlic	2 tsp
1/3 cup	minced red bell pepper	85 ml
1/2 cup	minced celery	125 ml
1/2 cup	dry sherry or white wine	125 ml
2 cups	long-grain brown rice, washed and drained	500 ml
1/2 cup	wild rice, washed and drained	125 ml
1/2 cup	white basmati rice, washed and drained	125 ml
4 cups	chicken stock	900 ml
1/2 tsp	dried thyme	1/2 tsp
1/4 tsp	dried sage	1/4 tsp
1 tbl	soy sauce or tamari	1 tbl

1. In a heavy pot over medium-high heat, sauté onion, garlic, red bell pepper, and celery in sherry until vegetables are soft (5–10 minutes).
2. Add rices and cook, stirring, for 3 minutes. Add stock, thyme, and sage, and bring to a boil. Lower heat to medium and cook, uncovered, for 15 minutes.
3. Lower heat to low, cover pot, and let pilaf steam until rice is tender (about 25 minutes). Stir in soy sauce and serve.

Serves 6.

Pilaf Pignoli

Pine nuts (pignoli) give this versatile rice dish authentic Mediterranean flavor.

2 tbl	unsalted butter	2 tbl
2 tbl	pine nuts	2 tbl
½	onion, chopped	½
¾ cup	long-grain rice, washed and drained	175 ml
½ tsp	salt (optional)	½ tsp
2 cups	chicken stock, boiling	500 ml

1. Melt 1 tablespoon of the butter in a heavy-bottomed saucepan. Add pine nuts and sauté until they turn light brown. Remove with a slotted spoon and reserve. Add onion to pan and sauté until soft and starting to turn golden. Remove and combine with pine nuts.
2. Melt remaining butter in pan. Add rice and sauté until golden. Return nuts and onion to pan.
3. Add salt, if used, to stock and pour over rice mixture. Bring to a boil. Cover and simmer very slowly until rice is tender and liquid has been completely absorbed (about 20 minutes). Remove pan from heat and allow to sit, covered, about 5 minutes. Fluff with a fork before serving.

Serves 4.

Dry-Roasting Rice

Toasting rice before cooking brings out the natural flavor of the grain and also reduces the cooking time. Wash and drain the rice as usual, then place it in a dry skillet over medium heat. Stir constantly until lightly browned (8–10 minutes). Then add water or other liquid to the same pan and cook as usual, decreasing the total cooking time by 10–12 minutes.

Eight-Treasure Rice Stuffing

Use this Asian-style stuffing for vegetables, fish, or poultry.

12	dried chestnuts	12
½ tsp	baking soda	½ tsp
2 cups	boiling water	500 ml
6	dried shiitake mushrooms, soaked and drained	6
1 cup	glutinous rice, washed and drained	250 ml
1¼ cups	chicken stock	300 ml
1 tbl	soy sauce	1 tbl
2 tsp	Asian sesame oil	2 tsp
2	Chinese sausages, thinly sliced	2
1 tbl	peanut oil	1 tbl
6	Chinese water chestnuts, peeled and diced	6
2	green onions, chopped	2
2 tbl	coarsely chopped cilantro (coriander leaves)	2 tbl

1. In a small bowl combine dried chestnuts, baking soda, and boiling water; soak 1 hour. Rinse and discard red skins. In a small saucepan bring to a boil 2 more cups (500 ml) water and add soaked chestnuts. Reduce heat and simmer 1 hour. Cut chestnuts into quarters; set aside.

2. Dice mushrooms and set aside.

3. In a heat-resistant dish, combine rice, stock, soy sauce, sesame oil, and sausages. Set aside 1 hour or overnight in refrigerator. Set bowl with rice mixture in steamer, cover, and steam over medium-high heat 30 minutes. Remove cover and let rice stand 15 minutes.

4. Place wok over medium-high heat and add peanut oil. Add reserved mushrooms, water chestnuts, soaked chestnuts, and green onions; stir-fry 1 minute and stir in reserved rice and cilantro.

Makes about 3 cups (700 ml).

Almond Rice with Chard

This recipe uses a variation on the Chinese Method of cooking rice (see page 10).

1 cup	long-grain white rice, washed and drained	250 ml
1 bunch	chard leaves	1 bunch
4 tbl	butter	4 tbl
3 cloves	garlic, minced	3 cloves
1 cup	fresh shelled peas, parboiled 2 minutes and drained	250 ml
to taste	salt and freshly ground black pepper	to taste
¼ cup	slivered almonds, toasted	60 ml

1. To a large pot of boiling water over high heat, add rice. Stir once, and boil, uncovered, until tender (about 15 minutes). Drain, rinse with cold water, and drain in strainer for 5 minutes.

2. Stack chard leaves, halve them lengthwise, and cut crosswise in strips ½ inch (1.25 cm) wide. Melt 2 tablespoons of the butter in a large skillet over medium-high heat. Stir in chard leaves and cook for 1 minute, or until leaves wilt. Reduce heat to low, add garlic, and cook until chard leaves are tender (about 2 minutes). Add peas, rice, and a dash of salt and pepper. Heat mixture over low heat, tossing lightly with a fork, until hot.

3. Add remaining 2 tablespoons butter. Cover pan and let stand until butter melts (about 2 minutes). Toss again lightly. Taste and add more salt and pepper if needed. Sprinkle with toasted almonds and serve hot.

Serves 4.

ARROZ DE MEXICO

This is a simplified version of a traditional Mexican sopa seca (dry soup) made with rice (arroz).

3 tbl	olive oil	3 tbl
1 cup	long-grain rice, washed and drained	250 ml
½	onion, chopped	½
1 clove	garlic, minced	1 clove
1 cup	mild green chile salsa	250 ml
2 cups	chicken stock	500 ml
as needed	fresh cilantro (coriander leaves), for garnish	as needed

1. Heat the oil in a skillet over medium heat. Add rice and cook, stirring constantly, until puffed and golden.
2. Push the rice to one side, add onion and garlic, and cook until onion is soft. Add salsa and cook briefly, stirring to blend rice with onion-salsa mixture.
3. Add chicken stock and bring to a boil. Reduce heat, cover, and simmer until all liquid has been absorbed (about 30–35 minutes). Garnish with cilantro and serve.

Serves 4.

PREPARING RICE-BASED SOPA SECAS

Sopa secas are a type of Mexican pilaf made with rice or sometimes vermicelli or corn tortillas. These "soupless soups" are made by sauteing rice until puffy and perfectly browned, then simmering in stock or other liquid with onions, garlic, and other seasonings. If you want to duplicate the delicate flavor and unique texture of an authentic rice sopa seca, use a regular long-grain variety—not converted or quick-cooking rice.

ARROZ VERDE

The piquant flavors of chiles and cilantro make this rice dish a favorite accompaniment for enchiladas, tacos, or fajitas.

2	Anaheim or poblano chiles, roasted, peeled, and seeded	2
½ cup	fresh cilantro (coriander leaves)	125 ml
2½ cups	chicken stock	600 ml
3 tbl	olive oil	3 tbl
1 cup	long-grain rice, washed and drained	250 ml
½	onion, chopped	½
1 clove	garlic, minced	1 clove
to taste	salt	to taste

1. Place chiles and cilantro in a blender or food processor with ½ cup (125 ml) of the chicken stock. Blend to a smooth purée and set aside.
2. Heat the oil in a skillet over medium heat. Add rice and cook, stirring constantly, until puffed and golden.
3. Push the rice to one side, add onion and garlic, and cook until onion is tender.
4. Add purée and remaining chicken stock and bring to a boil. Reduce heat, cover, and simmer until all liquid has been absorbed (about 30–35 minutes). Add salt to taste. Toss lightly with a fork before serving.

Serves 4.

Arroz Gualdo

Colorful and delicately flavored, this rice dish is typical of the cooking of Yucatán. Achiote, the small, brick-red seed of the annatto tree, imparts a bright yellow color to the rice and gives it a flavor that is especially tasty with Mexican-style fish dishes. Look for achiote in the ethnic foods section of supermarkets or in Mexican or Asian markets.

1½ tsp	achiote (annatto seed)	1½ tsp
3 tbl	olive oil	3 tbl
1 cup	long-grain rice, washed and drained	250 ml
½	onion, chopped	½
1 clove	garlic, minced	1 clove
2 cups	chicken stock or water	500 ml
1 tsp	salt	1 tsp

1. In a medium saucepan over low heat, sauté achiote in oil. Remove the seeds from the oil when they are dark brown and discard seeds. The oil will be dark orange in color.

2. Add the rice to the oil and sauté for about 5 minutes. Add the onion and garlic and cook until onion is soft.

3. In a separate pan bring stock to a boil. Add boiling stock to rice along with salt. Bring mixture to a boil, lower heat, cover, and cook until liquid is absorbed (20–25 minutes).

Serves 4.

Rice Soubise

A seasoned purée of onions and rice, a soubise is a fine accompaniment to grilled meats and poultry.

2 lb	white onions	900 g
5 tbl	unsalted butter	5 tbl
¾ cup	medium- or long-grain rice, washed and drained	175 ml
2 cups	fish or chicken stock	500 ml
1 tsp	salt	1 tsp
⅓ tsp	white pepper	⅓ tsp
⅓ cup	whipping cream	85 ml
2 tbl	unsalted butter	2 tbl

1. Peel and coarsely slice onions. Melt 3 tablespoons of the butter in pan, add onions, and cook for 5–7 minutes. Add rice, mix well, then add stock, salt, and pepper. Cover and simmer 30 minutes.

2. Purée the mixture in a food processor, adding cream and the remaining butter. Reheat over low heat. Adjust seasonings and serve.

Serves 4.

Risotto al Limone

Serve this sprightly Milanese risotto as a prelude to a fish or seafood main course (see cover photo). See Preparing Risotto on page 85.

3 tbl	unsalted butter	3 tbl
2 tbl	olive oil	2 tbl
¼ cup	minced onion	60 ml
2 tsp	grated lemon zest	2 tsp
1½ cups	Arborio rice, washed and drained	350 ml
4½ cups	heated chicken stock	1 l
¼ cup	lemon juice	60 ml
½ cup	freshly grated Parmesan cheese	125 ml
to taste	salt and freshly ground black pepper	to taste

1. In a heavy saucepan over moderately low heat, melt 2 tablespoons of the butter with the olive oil. Add onion and lemon zest; sauté slowly for 5 minutes. Add rice; stir to coat with oil. Turn up heat to high; toast rice, stirring, for 30 seconds. Immediately add ½ cup (125 ml) of the stock; reduce heat to medium low and stir until stock is absorbed.

2. Add more stock, ½ cup (125 ml) at a time, stirring constantly and adding more only when previous portion has been absorbed. When all stock is absorbed (about 20–25 minutes), stir in lemon juice. The rice should be tender with a firm center.

3. Stir in Parmesan cheese and remaining butter. Cook over low heat to blend and melt cheese. Season to taste with salt and pepper. Serve immediately in warm bowls.

Serves 4.

PREPARING RISOTTO

Arborio rice (traditionally used for risottos) can absorb several times its weight in liquid (usually stock), rendering it tender and creamy, with a firm center. Unlike pilaf, a risotto is stirred continuously over low heat. Risotto needs constant attention during the 20 or so minutes it takes to cook and should be served as soon as it is finished. For specific ingredients and cooking times, refer to the recipe you are using.

1. In a heavy saucepan heat oil or butter. Sauté seasonings such as minced onion. Add Arborio rice and stir to thoroughly coat grains.

2. Add ½–1 cup (125–250 ml) hot stock, stirring until stock is absorbed. Continue adding stock gradually, waiting until each addition has been absorbed before adding the next. Stir continuously to keep the rice from sticking.

3. When all hot stock is absorbed (20–25 minutes), rice will be tender and creamy, with a slight chewiness in the center of the kernels. Stir in flavoring, such as freshly grated Parmesan cheese or cream. Season to taste with salt and pepper. Serve immediately.

Risi e Bisi

The first spring peas prompt Venetian cooks to make this popular vegetable dish, which traditionally is souplike—thin enough to require a bowl and spoon.

1 tbl	olive oil	1 tbl
4 tbl	unsalted butter	4 tbl
3 tbl	minced shallot	3 tbl
1 cup	Arborio rice, washed and drained	250 ml
3 cups	chicken stock	700 ml
¾ lb	fresh shelled peas	350 g
6 tbl	freshly grated Parmesan cheese	6 tbl
3 tbl	minced fresh basil	3 tbl
1 tbl	minced parsley	1 tbl
to taste	salt	to taste

1. In a saucepan over moderately low heat, combine olive oil and 3 tablespoons of the butter. Add shallot; sauté gently until softened (about 5 minutes).

2. Add rice; cook, stirring, for 1 minute. Add stock, bring to a boil, cover, and reduce heat to low. Cook 2 minutes. Add peas and remaining butter. Cook until rice and peas are just tender (about 8–10 minutes). Stir in Parmesan, basil, and parsley. Add salt to taste. Serve immediately.

Serves 4.

Side Dishes for All Seasons

Steamed or sautéed fresh seasonal vegetables tossed with cooked rice make a quick side dish that will have everyone asking for seconds. Try asparagus, artichokes, young spinach or chard leaves, petit pois, snow peas, baby carrots, tiny green beans, finger-sized zucchini or summer squash, fresh limas, broccoli, kale or other leafy greens—whatever is in season—and don't forget fresh herbs.

Autumn Risotto

A trio of late-harvest vegetables—eggplant, celery, and mushrooms—added to a traditional risotto (see page 84) yields a hearty side dish for a simple autumn dinner.

½ cup	olive oil	125 ml
2 tbl	unsalted butter	2 tbl
¼ cup	minced onion	60 ml
2 tbl	minced garlic	2 tbl
1	eggplant, peeled and diced	1
½ tsp	hot-pepper flakes	½ tsp
½ cup	minced celery	125 ml
1 cup	cooked garbanzo or kidney beans (rinsed and drained, if canned)	250 ml
½ cup	quartered mushrooms	125 ml
1 tsp	minced fresh rosemary (optional)	1 tsp
¼ cup	minced parsley	60 ml
1 recipe	Risotto al Limone (see page 84)	1 recipe
to taste	salt and freshly ground black pepper	to taste
2 tbl	freshly grated Parmesan cheese	2 tbl

1. Heat oil and butter in a large skillet over moderate heat. Add onion and garlic and sauté until fragrant and slightly softened (about 3 minutes). Add eggplant and brown on all sides. Add hot-pepper flakes and celery and cook an additional 3 minutes. Remove from heat.

2. Preheat oven to 375°F (190°C). Stir beans, mushrooms, rosemary (if used), and half the parsley into eggplant mixture. Combine vegetables and prepared risotto. Season to taste with salt and pepper. Transfer to a buttered 2-quart (1.8-l) casserole, cover, and bake until heated through (about 20 minutes). Serve from casserole, topped with remaining parsley and Parmesan cheese.

Serves 6.

Céleri-Rave et Ris

Try serving this smooth purée of celery root (celeriac) and rice with grilled meats or baked ham. With a subtle, celerylike flavor that complements rice in side dishes such as this one, celery root is well worth seeking out at farmers markets and supermarkets. You can make the recipe up to 6 hours ahead. Reheat the purée slowly in a double boiler, stirring well, and serve hot.

1	celery root, peeled and cut into 1-inch (2.5-cm) dice	1
1 cup	half-and-half	250 ml
4 cups	chicken stock	900 ml
⅔ cup	long-grain white rice, washed and drained	150 ml
2 tbl	unsalted butter	2 tbl
2 tbl	sour cream	2 tbl
to taste	salt and freshly ground black pepper	to taste

1. In a large saucepan over high heat, bring celery root, half-and-half, and stock to a boil; reduce heat to maintain a simmer. Cover and cook 10 minutes. Add rice and cook, uncovered, until celery root and rice are tender (about 20 minutes). Drain, reserving liquid.

2. Transfer contents of saucepan to a food processor or blender and purée. Return to saucepan. Add butter, sour cream, and just enough of the reserved cooking liquid to make a thick purée. Season to taste with salt and pepper. Serve at once or let cool, refrigerate, and reheat before serving.

Serves 6.

Something Sweet

Combine rice with fruits, sweeteners, and spices, and you can reap sweet success. For a luscious new twist on breakfast or dessert, serve rice-flour waffles with a trio of sweet sauces, or start off the perfect day (or end it) with fragrant, soul-warming rice pudding or almond-scented torta riso.

Rice Waffles with a Trio of Fruit Sauces

Light-textured rice waffles and a selection of luscious sauces make an unusual brunch or dessert. Each of the sauce recipes makes about 2 cups (500 ml).

Strawberry Sauce

2 cups	sliced strawberries	500 ml
½ cup	sugar	125 ml

Piña Colada Sauce

12 oz	piña colada mix	375 ml
1 tbl	cornstarch	1 tbl
2 tbl	rum	2 tbl
½ cup	flaked coconut	125 ml

Hot Fudge Sauce

1 can (14 oz)	sweetened condensed milk	1 can (425 ml)
4 squares (1 oz each)	unsweetened chocolate	4 squares (30 g each)
1 tbl	butter	1 tbl
1 tsp	vanilla extract	1 tsp

as needed	oil, for coating waffle iron	as needed
¾ cup	sifted brown rice flour	175 ml
2 tsp	baking powder	2 tsp
1 tsp	ground nutmeg	1 tsp
1 tsp	ground cinnamon	1 tsp
pinch	salt	pinch
1 tbl	honey	1 tbl
2	eggs, separated	2
1 cup	nonfat milk	250 ml
¼ cup	oil	60 ml
1 cup	cooked rice	250 ml

1. To prepare Strawberry Sauce, in a small saucepan combine strawberries and sugar; let stand 1 hour. Over medium heat cook 5 minutes, mashing berries and adding 2–4 tablespoons water, a tablespoon at a time, to achieve desired consistency. Purée in blender; serve hot or cold.

2. To prepare Piña Colada Sauce, pour piña colada mix into a medium saucepan over high heat and bring to a boil. Combine cornstarch with 1 teaspoon water and stir rapidly into hot mix. Cook until slightly thickened (about 3–4 minutes). Remove from heat and stir in rum and coconut. Let cool slightly before serving.

3. To prepare Hot Fudge Sauce, in a medium saucepan over medium heat, combine condensed milk and chocolate and heat until chocolate is nearly melted (about 3–4 minutes), stirring often. Stir in butter and vanilla until mixture is smooth and thick. Serve warm.

4. To prepare waffles, preheat waffle iron and lightly brush with oil. In a medium bowl combine rice flour, baking powder, nutmeg, cinnamon, and salt. In a separate bowl combine honey, egg yolks, milk, oil, and rice. Beat egg whites until stiff peaks form. Combine dry and wet ingredients, then fold in egg whites.

5. Preheat oven to 200°F (95°C). Cook waffles, one at a time, keeping them warm in the oven on a heatproof platter. Serve with sauces.

Makes about 8 waffles.

Torta di Riso

The same Arborio rice that makes a creamy risotto can create a luscious baked pudding.

½ cup	almond-flavored liqueur	125 ml
½ cup each	golden raisins, dried currants, minced dried figs	125 ml each
½ cup	Arborio rice, washed and drained	125 ml
4 cups	milk	900 ml
1½ tsp each	grated lemon and orange zest	1½ tsp each
¾ cup	sugar	¾ cup
6	eggs	6
¾ cup	chopped almonds	175 ml
1 tsp	almond extract	1 tsp
as needed	butter and sugar, for coating pan	as needed

1. Combine liqueur, ¼ cup (60 ml) of water, raisins, currants, and figs in a small saucepan. Bring to a simmer over medium heat and let simmer 5 minutes. Set aside 2–3 hours until all liquid is absorbed.

2. In a saucepan over high heat, bring 4 cups (900 ml) water to a boil. Add rice and cook 3 minutes. Drain, then return rice to a clean saucepan with milk and lemon and orange zest. Bring to a simmer over moderately high heat, then reduce heat to low, cover, and cook 1 hour. Remove from heat, cool, and add ¾ cup (175 ml) sugar. Add eggs one at a time, then stir in nuts, almond extract, and fruit mixture.

3. Preheat oven to 325°F (160°C). Butter bottom and sides of a 9-inch (25-cm) round cake pan, then coat with sugar, shaking out excess. Pour batter into prepared pan. Bake until a knife inserted in center comes out clean (about 1 hour). Remove torta to a rack; cool in pan. To serve, turn torta out of pan and cut into wedges. Serve warm or cold.

Serves 8.

Orange Rice Pudding

This version of a traditionally rich dessert has been updated with a hint of citrus and streamlined to reduce the amount of fat.

4 cups	nonfat milk	900 ml
2 tbl	maple syrup	2 tbl
pinch	salt	pinch
1 cup	long-grain brown rice, washed and drained	250 ml
1½ tsp	vanilla extract	1½ tsp
1 cup	raisins	250 ml
⅓ cup	chopped, pitted dates	85 ml
1 tbl	grated orange zest	1 tbl
½ cup	orange juice	125 ml
as needed	oil, for coating baking dish	as needed
as needed	orange zest strips, for garnish	as needed

1. In a heavy saucepan combine 2 cups (500 ml) of the milk, the maple syrup, salt, and rice. Bring to a boil and lower heat to simmer. Cook 25 minutes over medium heat.

2. Preheat oven to 325°F (160°C). In separate bowl, mix together the remaining ingredients except oil and zest for garnish. Lightly oil a 9- by 12-inch (22.5- by 30-cm) baking dish.

3. Mix rice mixture with orange juice mixture and pour into the prepared baking dish. Bake until solidified and lightly browned (about 90 minutes). Garnish with strips of orange zest.

Serves 8.

Index

Note: Page numbers in italics refer to photos.

Almond Rice with Chard 78, *79*
Appetizers
  California Roll Sushi 18, *19*
  Dolmas 24
  Nigiri Sushi 21
  Pearl Balls with Pine Nut-Spinach Filling *14*, 15
  Sizzling Rice Cakes 16, *17*
  Suppli al Telefono 25
  Sushi Rice 20
Arroz de Mexico 80
Arroz Gualdo 82
Arroz Verde 81
Autumn Risotto 87
Avgolemono 30, *31*

Basque Paella 54, *55*
Beef and pork
  Bon Temps Jambalaya 44, *45*
  Brown Jambalaya 46
  Eight-Treasure Rice Stuffing 77
  Hoppin' John 42, *43*
  Indonesian Fried Rice 64, *65*
  New Orleans Red Beans and Rice *40*, 41
  Pearl Balls with Pine Nut-Spinach Filling *14*, 15
  Rice Pilaf with Parmesan Cream *70*, 71
  Riz à la Basquaise 60
  Sausage and Fennel Fried Rice 62, *63*
  Seafood Filé Gumbo *12*, 32
  Shrimp and Pork Fried Rice 61
  Shrimp Okra Pilau 48, *49*
  Suppli al Telefono 25
  Szechuan Spareribs with Toasted Rice 56, *57*
Bon Temps Jambalaya 44, *45*
Breton Mussel and Shrimp Soup with Rice *26*, 27
Brown Jambalaya 46

California Roll Sushi 18, *19*
Céleri-Rave et Ris 88, *89*
Chicken
  and Rice in a Clay Pot 58, *59*
  Arroz de Mexico 80
  Arroz Gualdo 82
  Arroz Verde 81
  Autumn Risotto 87
  Avgolemono 30, *31*
  Basque Paella 54, *55*
  Brown Jambalaya 46
  Céleri-Rave et Ris 88, *89*
  Chicken-Rice Soup 28
  Dolmas 24
  Eight-Treasure Rice Stuffing 77
  Kedgeree 68, *69*
  Lentil-Pilaf Salad 36, *37*
  Pilaf Minceur 74, *75*
  Pilaf Nouvelle 72, *73*
  Pilaf Pignoli 76
  Rice Soubise 83
  Risi e Bisi 86
  Risotto al Limone 84
  Riz à la Basquaise 60
  Roast Chicken with Pilaf Stuffing 50, *51*
  Sausage and Fennel Fried Rice 62, *63*
  Shrimp and Pork Fried Rice 61
  Sizzling Rice Soup 29
  Zuppa di Riso 33
Cooking methods
  Chinese Method 10
  Electric Rice Cooker Method 11
  Measured Water Method 9
  Microwave Method 10–11
  Neverfail Rice 22–23
  Preparing Risotto 85
  Unlimited Water Method 10

Desserts
  Orange Rice Pudding 94
  Rice Waffles with a Trio of Fruit Sauces *90*, 91
  Torta di Riso 93
Dolmas 24
Dry-Roasting Rice 76

Eight-Treasure Rice Stuffing 77

Fish and shellfish
- Basque Paella 54, *55*
- Bon Temps Jambalaya 44, *45*
- Breton Mussel and Shrimp Soup with Rice *26,* 27
- California Roll Sushi 18, *19*
- Four-Treasure Rice 67
- Indonesian Fried Rice 64, *65*
- Kedgeree 68, *69*
- Mandarin Fried Rice 66
- *Nigiri Sushi* 21
- Pearl Balls with Pine Nut-Spinach Filling *14,* 15
- Rice Salad Niçoise 38, *39*
- Rice Venetian 52
- Risotto with Smoked Salmon 53
- Seafood Filé Gumbo *12,* 32
- Shrimp and Pork Fried Rice 61
- Shrimp Okra Pilau 48, *49*

Four-Seasons Rice Salad 35
Four-Treasure Rice 67

Hoppin' John 42, *43*
Hot Fudge Sauce *90,* 91

Indonesian Fried Rice 64, *65*

Kedgeree 68, *69*

Lentil-Pilaf Salad 36, *37*

Mandarin Fried Rice 66
Minnesota Pilaf with Cashew Gravy 47

Neverfail Rice 22–23
New Orleans Red Beans and Rice *40,* 41
Nigiri Sushi 21

Off-the-Shelf Rice Salad 39
Orange Rice Pudding 94

Pearl Balls with Pine Nut-Spinach Filling *14,* 15
Pilaf Minceur 74, *75*
Pilaf Nouvelle 72, *73*
Pilaf Pignoli 76
Piña Colada Sauce *90,* 91
Preparing "Fried Rice" 61
Preparing Rice-Based Sopa Secas 80
Preparing Risotto 85

Rice
- and Sprouts Salad 34, *35*
- cooking 8–11
- *Dry-Roasting Rice* 76
- *for Risotto* 52
- *Neverfail Rice* 22–23
- Pilaf with Parmesan Cream *70,* 71
- *Preparing "Fried Rice" 61*
- *Preparing Risotto* 85
- Salad Niçoise 38, *39*
- Soubise 83
- storing and preparing 8
- Sushi Rice 20
- varieties 6–7
- Venetian 52
- Waffles with a Trio of Fruit Sauces *90,* 91

Risi e Bisi 86
Risotto al Limone 84
Risotto with Smoked Salmon 53
Riz à la Basquaise 60
Roast Chicken with Pilaf Stuffing 50, *51*

Sausage and Fennel Fried Rice 62, *63*
Seafood Filé Gumbo *12,* 32
Shrimp and Pork Fried Rice 61
Shrimp Okra Pilau 48, *49*
Side Dishes for All Seasons 86
Sizzling Rice Cakes 16, *17*
Sizzling Rice Soup 29
Soups and stews
- Avgolemono 30, *31*
- Breton Mussel and Shrimp Soup with Rice *26,* 27
- Chicken-Rice Soup 28
- Seafood Filé Gumbo *12,* 32
- Sizzling Rice Soup 29
- Zuppa di Riso 33

Stir-Fry Pantry, The 67
Strawberry Sauce *90,* 91
Suppli al Telefono 25
Sushi Rice 20
Szechuan Spareribs with Toasted Rice 56, *57*

Torta di Riso 93

Zuppa di Riso 33